A VERY Brief Guide to AQA GCSE Physics 2

A revision guide for those in a hurry.

First published December 2015
ISBN: 1522822976
ISBN-13: 978-1522822974

Foreword

AQA GCSE Physics is a popular exam course. It shares two units (P1 and P2) with the GCSE Science course. This book provides a quick look at the content of Unit P2.

The author has an MA in Natural Science and an MSci in Experimental and Theoretical Physics from the University of Cambridge. He has taught physics in state and independent schools for over 15 years. He is a lead examiner of physics A-level and an examiner of physics GCSE.

Introduction

<u>About This Guide</u>

This revision guide is split into topics in line with the AQA specification. All the key points are covered in a clear bullet point format. Material only assessed on higher tier papers is in red.

<u>Revision Technique</u>

Effective revision is that which starts early and is frequently added to. Whenever you finish a topic in lessons, you should revise it. Every month or so, you should take a look back at earlier topics and refresh your memory.

Effective revision is active. Reading through text is a good start, but you are unlikely to remember everything. Instead of passively reading, you should write things down, for instance on cards or on a poster. One of the best things to do is attempting questions – you can download past papers from the AQA website (http://www.aqa.org.uk/subjects/science/gcse/physics-4403/past-papers-and-mark-schemes) or buy my Physics Problems for GCSE book (http://www.archaeoroutes.co.uk/edphys/problems.php).

Using This Revision Guide

This book lends itself particularly well to use throughout the course. However, it is also eminently suitable to use for revision in the run up to the exam.

Here are a few ideas of ways you can use it:

1. Read a topic and write it out in your own words.
2. Get together with friends, all read the same topic and discuss it.
3. Work through a topic and annotate or correct your class notes with things you missed or wrote down wrong.
4. After a lesson, find the appropriate section in the guide and check that you have made complete notes.
5. Look up anything you don't know, for instance when doing homework.
6. Get a friend or family member to invent questions and then test you. After a while swap over – writing questions is as good a revision technique as answering them.
7. Go through a topic and traffic light it. This means colouring each bullet point green (if you are happy), yellow (if you are a bit unsure) or red (if you have no idea) so you can target your revision where you need it most. *If you are reading this as an ebook, your reader probably has highlighting tools you could use.*
8. Teach a lesson on a topic to a friend or family member. They should have the guide in front of them to tick things off as you go through them. They can then give you feedback.

Another really good use of this revision guide is as a focus for tutorials. You could work through a topic and then ask your tutor/teacher about anything that you didn't understand. Write your questions down as they come up or you might forget them!

Tiers

There are two levels of exams for this course. You might be entered for the higher tier exam or the foundation tier exam.

Mostly the difference is in the language used to ask the questions, and the level of maths required. However, there are also differenced in the content that could be examined.

Throughout this guide I have highlighted material that is only relevant for higher tier candidate like this.

A Quick Note about Maths

There are 13 (14 for higher tier) formulæ that are given on the formulæ sheet (in letter form) and you are expected to use:

- $$acceleration \ (m/s^2) = \frac{force \ (N)}{mass \ (kg)}$$

  or $\quad force \ (N) = mass(kg) \times acceleration \ (m/s^2)$

- $$acceleration \ (m/s^2) = \frac{change \ in \ velocity \ (m/s)}{time \ taken \ (s)}$$

- $weight \ (N) = mass \ (kg)$
  $\times gravitational \ field \ strength(N/kg)$

- $force \ (N) = spring \ constant(N/m) \times extension \ (m)$

- $work \ done \ (J) = force \ (N) \times distance \ moved \ (m)$

- $$power \ (W) = \frac{energy \ transferred \ (J)}{time \ taken \ (s)}$$

- change in GPE (J) = mass (kg) x
  gravitational field strength (N/kg)
  x change in height (m)

- KE (J) = ½ x mass (kg) x speed(m/s) ^{2}

- momentum (kg m/s) = mass (kg) x velocity (m/s)

- $$current \ (A) = \frac{charge \ (C)}{time \ (s)}$$

- $$potential \ difference \ (V) = \frac{work \ done \ (J)}{charge \ (C)}$$

- potential difference (V) = current (A) x resistance (Ω)

- $power\ (W) = voltage\ (V)\ x\ current\ (A)$

- $energy\ transferred\ (J) = potential\ difference\ (V)$ $\times charge\ (C)$

There are also a couple of others that could be useful:

- stopping distance = thinking distance + braking distance

- $$frequency\ (Hz) = \frac{1}{time\ period\ (s)}$$

Foundation tier papers always give the numbers in such a way that the formulæ can be used in this form. Higher tier papers are likely to need them rearranging before use – remember how to rearrange by doing the same thing to both sides of the equation.
Higher tier candidates also have to watch units. Some values will not be given in the correct units (eg. cm instead of m) and will need to be converted. Foundation tier candidates are spared this extra step!

A Quick Note about How Science Works

Scientists look at evidence. This comes from experiments and observations of the world. They then devise models or theories about what causes these things to happen. A good theory fits the current observations AND makes new predictions that can be tested.

Old theories get replaced by newer ones IF the new one is simpler or if it fits new evidence that the old one doesn't.

If you are asked to carry out an experiment to test a theory, you need to consider the following things:

- Is your sample size large enough?

    If not you may just be seeing a random effect. If three out of four tests of a new car tyre showed it reduced the braking distance, they might just be three sets of tyres that were particularly well made.

- Have you eliminated the effects of other factors?

    If not they could be affecting your results. Perhaps the tyres don't work well on wet roads.

- Have you eliminated any bias?

    If not you might not accept the evidence. If the company making the tyres is paying you to test their product, you might be worried about not getting hired again if you say it doesn't work.

A classic example of poor evidence is that given in shampoo adverts. For instance one claims 90% of those asked said their hair was silkier with the brand. Initially that sounds impressive.

Looking at the small print you see that the company itself asked 30 of their repeat customers. This is far too small a sample and is also biased.

In reality, it is sometimes unreasonable to carry out an experiment. This could be because:

- It would be too expensive.
- It would be unethical, for instance an experiment that could harm people.

P2.1.1 Resultant Forces

Forces are how things interact. They make things change. To make something happen you exert a force. Forces can either help each other or work against each other.

'Resultant force' is the name we give to a single force that would do the same job as all the 'real' forces acting on something.

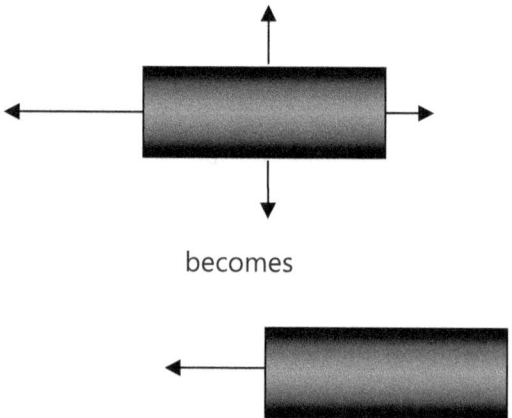

becomes

The resultant force is the sum of the all forces acting on an object.

- If the forces are in the same direction, you need to add them up.

3N

6N ←————————• resultant = 3N + 6N
= 9N to the left

- If the forces are in opposite directions, you need to subtract the smaller from the larger.

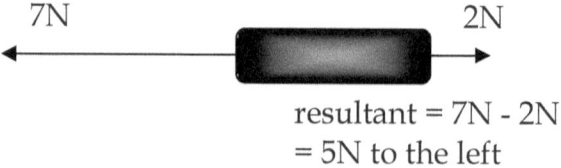

resultant = 7N - 2N
= 5N to the left

- If the forces in each direction are the same size there is no resultant force. We say the forces are balanced.

When a resultant force acts on an object, it makes it move differently.

- It accelerates in the direction of the force.
- Accelerating could mean speeding up, slowing down or changing direction, depending on which way the force is pointing.
- We will meet the formula that quantifies the acceleration in P2.1.2.

When there isn't a resultant force (ie. it is zero) the object's movement doesn't change.

- If it is stationary (not moving), it will remain stationary.
- If it is moving, it will continue to move in exactly the same way (speed and direction).

You can use this in reverse too. That is, if you know an object does not move you know there is zero resultant force acting upon it.

When two objects interact with each other, the forces they apply to each other are the same size and in opposite directions.

- "Every action has an equal and opposite reaction." – Sir Isaac Newton
- We call them a 'pair' of forces. Each one in the pair is exerted on a different object. This means that they don't cancel each other out.

P2.1.2 Forces and Motion

In P2.1.1 we saw that unbalanced forces cause things to move differently. Now we will look at giving numbers to those changes. We call that a quantitative analysis, as opposed to a qualitative one.

Isaac Newton was able to relate the acceleration of an object to its mass and the force applied to it.

- If the resultant force on an object got bigger, the acceleration of the object got bigger.
- If the object's mass increased, the acceleration for a given force increased.

A car with a low mass and large resultant force will have a high acceleration. That's why car companies go to long lengths to make engines more powerful, but also make the car lighter by using materials like aluminium and carbon fibre.

Cue possibly one of the most famous physics formulæ:

$$acceleration\ (m/s^2) = \frac{force\ (N)}{mass\ (kg)} \qquad a = {^F}/_m$$

It is more commonly written as:

$$force\ (N) = mass(kg) \times acceleration\ (m/s^2)$$
$$F = m \times a$$

Using graphs to represent motion allows us to see changes more easily. A distance-time graph is a graph of distance (on the y-axis) against time (on the x-axis).

- Remember that $$speed\ (m/s) = \frac{distance\ (m)}{time\ (s)}$$ $v = {}^{d}/_{t}$
- The gradient of a distance-time graph is thus the speed.

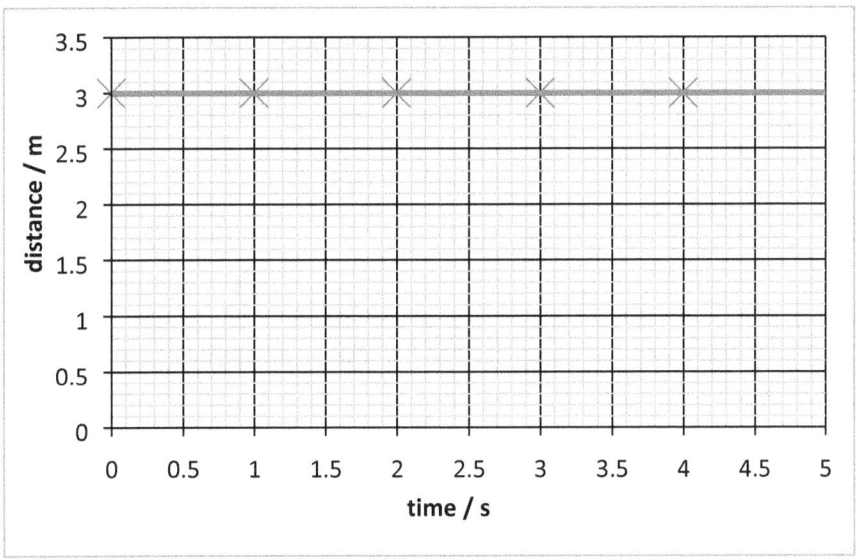

In the graph above, the object is not moving.

You can tell it isn't moving because it stays the same distance away the whole time.

OK, it could be going in a circle around the point from which we are measuring, but that isn't going to come up.

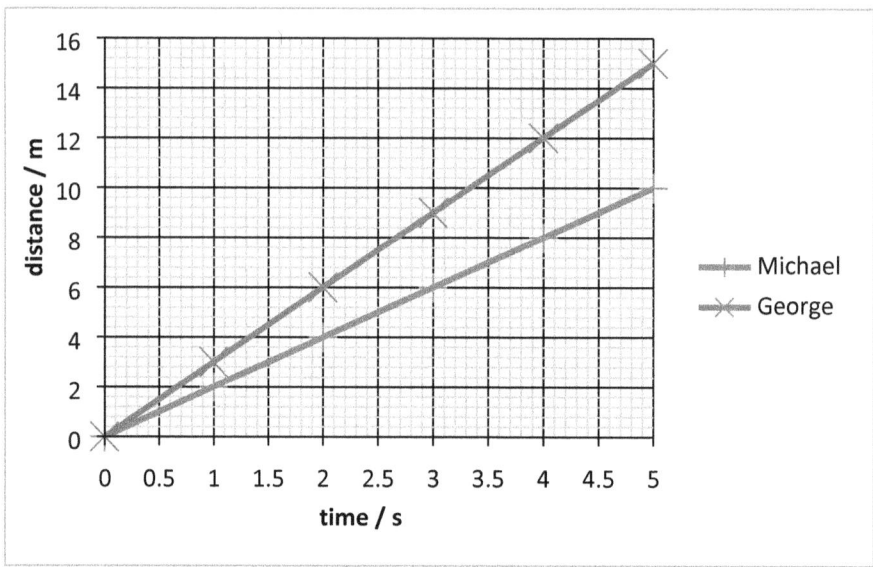

In the graph above, George is moving faster than Michael.
You can tell that because George's line is steeper (has a higher gradient).

George is moving at:

$$\frac{change\ in\ y}{change\ in\ x} = \frac{15 - 0}{5} = 3m/s$$

'Velocity' is another word we use to describe motion. It carries more information about how something is moving than 'speed'.
- Velocity is defined as speed and direction of travel.
- Eg. 8m/s is a speed; 25m/s due east is a velocity
- We can simply work out change in velocity by subtracting the initial velocity from the final velocity.   a=v-u

The acceleration of an object is how much it speeds up, or slows down, in a given amount of time.

- Acceleration can be defined as 'rate of change of velocity'.
- $$acceleration\ (m/s^2) = \frac{change\ in\ velocity\ (m/s)}{time\ taken\ (s)}$$

$$a = \frac{v - u}{t}$$

Eg. A hockey player on a training session accelerates from a jog of 3.0m/s to a sprint of 6.0m/s in 2.0s. Her acceleration is:

$$\frac{6.0 - 3.0}{2.0} = 1.5m/s^2$$

A velocity-time graph is a graph of velocity (on the y-axis) against time (on the x-axis).

- The gradient of a velocity-time graph gives the acceleration.

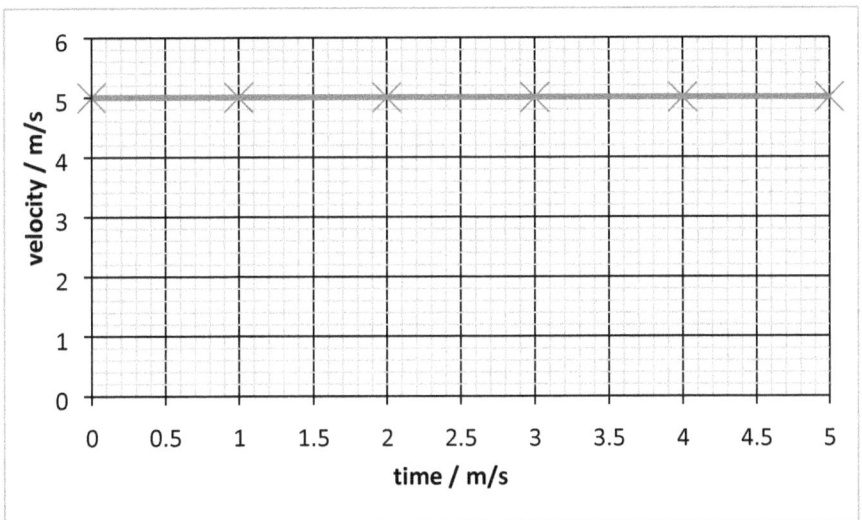

In the graph above, the object is moving at a constant velocity of 5m/s.

You can tell it is a constant velocity because the line is horizontal.

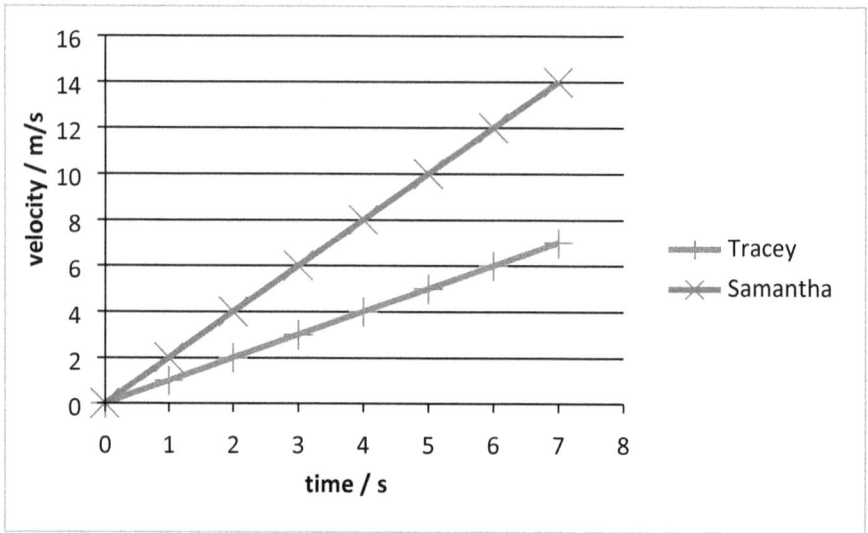

In the graph above, Samantha has a higher acceleration than Tracey.

You can tell that her acceleration is higher because the line is steeper (has a higher gradient).

Samantha has an acceleration of:
$$\frac{14 - 0}{7} = 2m/s^2$$

- The distance travelled by an object is shown by the area under the line on its velocity-time graph.

Samantha travels:
$$^1/_2 \times 7 \times 14 = 49m$$

P2.1.3 Forces and Braking

Slowing down can be just as important as speeding up. In fact lives can depend on it: a common real-world example is a car braking. The size of the braking force is only one part of working out how far a vehicle will travel before it comes to complete stop.

An object that is moving at a constant speed has no resultant force acting on it.
- The forces acting on it are balanced.
- We call the forward force the 'driving force'.
- We call the backward forces 'resistive forces'. Mostly these are air resistance and friction.

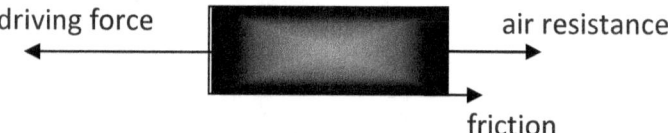

driving force   air resistance

friction

The faster a vehicle travels, the harder it has to brake in order to stop before hitting something. That means: the higher the speed, the bigger the braking force needed.

In an emergency stop, a driver would use the largest braking force the vehicle is capable of exerting (given the conditions).
- We call the distance required to stop in this situation the 'stopping distance'.
- The faster a car travels, the greater the stopping distance is.

We can break the stopping distance down into two parts:
- We call how far the car travels between the threat appearing and the driver starting to brake the 'thinking distance'.
- We call how far the car travels between starting to brake and coming to a complete stop the 'braking distance'.
- stopping distance = thinking distance + braking distance

Typical Stopping Distances

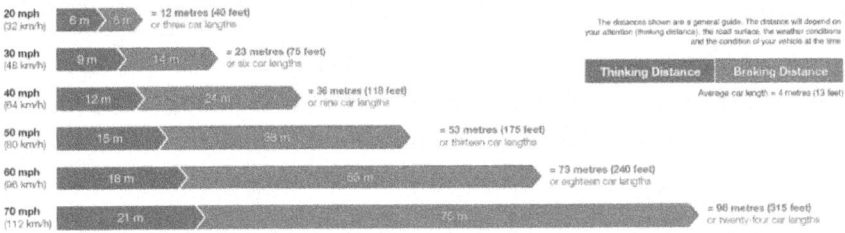

The distances shown are a general guide. The distance will depend on your attention (thinking distance), the road surface, the weather conditions and the condition of your vehicle at the time.

Thinking Distance	Braking Distance

Average car length = 4 metres (13 feet)

20 mph (32 km/h) — 6 m / 6 m = 12 metres (40 feet) or three car lengths

30 mph (48 km/h) — 9 m / 14 m = 23 metres (75 feet) or six car lengths

40 mph (64 km/h) — 12 m / 24 m = 36 metres (118 feet) or nine car lengths

50 mph (80 km/h) — 15 m / 38 m = 53 metres (175 feet) or thirteen car lengths

60 mph (96 km/h) — 18 m / 55 m = 73 metres (240 feet) or eighteen car lengths

70 mph (112 km/h) — 21 m / 75 m = 96 metres (315 feet) or twenty-four car lengths

The image above (from the Highway Code) shows how the thinking and braking distances increase with increasing speed of the car.

The thinking distance depends on the speed of the car and the driver's reactions.
- The faster the car, the further it will go while the driver reacts.
- The longer it takes the driver to react, the further the car will travel in that time.
- Remember distance = speed x time.

Many things can make a driver react slower, including:
- distraction (eg. using a mobile 'phone)
- tiredness
- alcohol
- drugs
- getting older

Many things can increase the braking distance, including:
- higher speeds
- more mass (eg. carrying more passengers)
- worn brakes or tyres
- adverse road conditions (eg. wet, icy, loose chippings)

The brakes do work.
- The friction between the brakes and the wheel converts kinetic energy to heat energy (and possibly a bit of sound).
- In car races at night you can often see them glowing red hot!

P2.1.4 Forces and Terminal Velocity

Things always have a top speed (unless they are in a vacuum). This is due to how resistive forces like air resistance work. This applies to things like cars, parachutists, cyclists, etc. Vast amounts of time and money is spent on making this top speed higher for athletes, cars etc.

We use the word 'fluid' to mean 'liquid or gas'.
Anything moving through a fluid experiences a resistive force.
- This resistive force is made up of drag and friction.
- A common example of this is air resistance.
- The faster an object moves, the larger the resistive force becomes.
- The larger the resistive force is, the harder it is for the object to accelerate.
- As there is always a maximum possible driving force, then there will always come a point where the resistive force balances the driving force.
- When this happens, there is no resultant force (and thus there is no acceleration).
- The object has reached its top speed. We call this its terminal velocity.

The size and shape of the moving object also affects the resistive forces acting upon it.
- The larger the object is, the larger the resistive force it experiences.
- A shape with low resistive force is called streamlined or aerodynamic.
- Typically, aerodynamic shapes have smooth curves and avoid flat surfaces at the front and the back.
- The air flows easily around the object and exerts less force on it as a result.

A common example you will encounter in questions is a parachute jump. In this case, the driving force is the weight of the parachutist and the resistive force they experience is the air resistance. So, the parachutist initially accelerates due to gravity, before eventually reaching a steady speed.

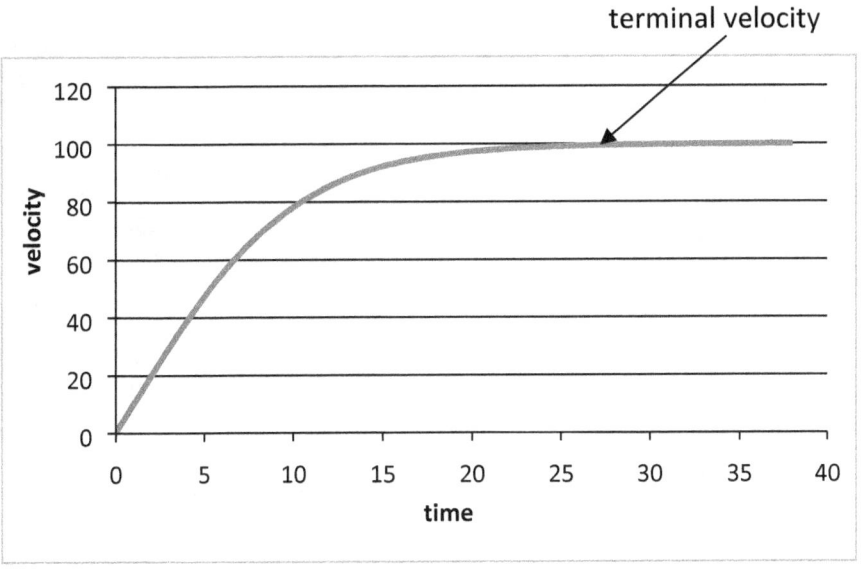

- When the parachutist opens his parachute, the effective area suddenly become much larger.
- This increases the resistive force.
- The upwards force is now larger than the downwards force, so he slows down.
- As he slows down, the resistive force decreases.
- The forces will soon balance again, and he will move at a new, slower terminal velocity.

The same principle affects many things.

- The top speed of a car is determined by the power of its engine and how well streamlined it is.
- How fast a cyclist can go depends on their strength, the shape they adopt and the clothes they wear.

Weight is the force exerted on an object due to a gravitational field.
- It is measured in newtons (N), like all forces.
- The more mass an object has, the greater its weight will be.
- The higher the gravitation field strength, the greater the weight will be.
- We often shorten 'gravitational field strength' to 'g'.
- On Earth we usually round the gravitational field strength to 10N/kg.

$$weight\ (N) = mass\ (kg) \times gravitational\ field\ strength(N/kg)$$
$$W = m \times g$$

Eg. An apple has a mass of 0.17kg.
The weight of the apple, on Earth, is: 0.17 x 10 = 1.7N

P2.1.5 Forces and Elasticity

Forces can also deform objects as well as move them around. Architects and engineers need to consider this when they are designing buildings, tools etc.

Forces are able to change the shape of objects.
- This change can be bending, stretching or squashing.
- These are collectively called 'deforming'.

Some materials are elastic. Some are plastic.
- A material that returns to its original shape and size when you let go is called elastic.
- If it doesn't return to its original shape and size when you let go, it is called plastic.
- Elastic and plastic are also names of specific materials, so be careful not to get confused.

As with most things, if you change something, you do work on it.
- When you deform an elastic object you are storing elastic potential energy in it.
- When you release the object, this elastic potential energy can be used to do work.

The amount an object is stretched is called its 'extension'.
- The extension is how much longer it is than its length when you started to stretch it.

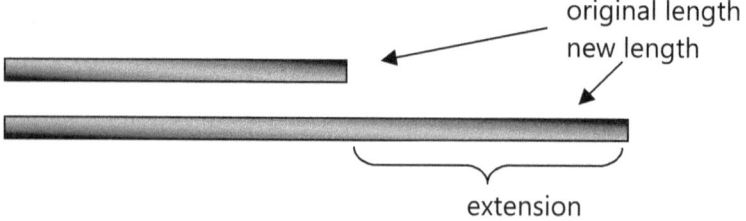

original length
new length
extension

Hooke's Law states that:
$$force\ (N) = spring\ constant(N/m) \times extension\ (m)$$
$$F = k \times e$$

- We call how far the spring stretches when it is pulled with 1N the 'spring constant'.
- If the spring constant is in N/m, the distance is in m. If it is in N/cm, the distance is in cm.

Eg. If something with a spring constant of 5N/cm is stretched 4cm it required a force of 5 x 4 = 20N.

It is often useful to plot a graph of extension (on the y axis) against force (on the x axis) to see how different materials behave.
- Once the graph starts to curve, it is no longer obeying Hooke's Law.
- We say that the 'limit of proportionality' has been passed.
- If your graph doesn't go through the origin (0,0) it is likely that you have plotted length not extension.

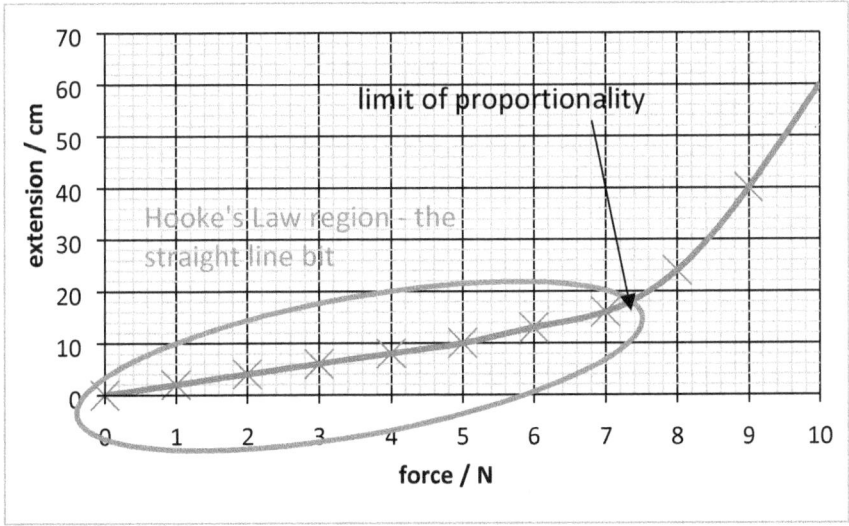

P2.2.1 Forces and Energy

Energy and power are both to do with work. This might be an alien concept to some people, but it is crucial to achieving anything! The difference between energy and power is just a matter of time (sorry, I'll cut the bad jokes for a bit).

When we use a force to move something we say we are 'doing work'.

- $work\ done\ (J) = force\ (N) \times distance\ moved\ (m)$

  $W = F \times d$

- 'Work done' is the same as 'energy used' or 'energy transferred'.

Eg. I use 12N to drag a sledge 10m. I do 12x10=120J of work.

The object might move at constant speed or accelerate.

- If the force is needed just to overcome friction, or other resistive forces, the forces are balanced and the object moves at constant speed.
- All the work goes into heat (and sound).
- If the force exerted is larger than that, the object will accelerate.
- In addition to heat (and sound) some of the work goes into kinetic energy.

We call the rate of doing work 'power'.

- Power is the amount of work done (or energy transferred) every second.

- $$power\ (W) = \frac{energy\ transferred\ (J)}{time\ taken\ (s)} \qquad P = {}^{E}/_{t}$$

- We measure power in watts.

Eg. If I do 500J of work in 10s I have a power of:

$$\frac{500}{10} = 50W$$

If we are lifting the object, then the force needs to be at least as large as the object's weight.
- work done (J) = weight (N) x distance lifted (m)
  work done (J) = mass (kg) x gravitational field strength (N/kg) x change in height (m)
- We say that the object has gained gravitational potential energy:
  change in GPE (J) = mass (kg) x gravitational field strength (N/kg) x change in height (m) $E_p = m \times g \times h$

Eg. I lift a 3kg object 5m on Earth. I give it 3 x 10 x 5 = 150J of GPE.

Remember, sometimes extra force is used to accelerate the object.
- In that case, the extra work done goes into the kinetic energy of the object.
- KE (J) = ½ x mass (kg) x speed(m/s) ^{2} $\qquad E_k = ½ \times m \times v^2$

Eg. A raven has a mass of 1.4kg and flies at 10m/s. Its kinetic energy is ½ x 1.4 x 10^2 = 70J.

When an object falls, it loses altitude.
- The object's GPE is converted into KE.
- GPE at top = KE at bottom (assuming no air resistance).
- If it does experience air resistance, then some of the kinetic energy gets converted into heat, due to the friction with the air. That is why things re-entering our atmosphere glow red hot.

P2.2.2 Momentum

Things that are moving tend to keep on moving. If you kick a ball, it doesn't stop the instant it leaves your foot. Stopping a train is a lot harder than stopping a bicycle. This is all due to momentum.

All moving objects have momentum.
- The more momentum an object has, the harder it is to stop it moving.
- momentum (kg m/s) = mass (kg) x velocity (m/s)

$$p = m \times v$$

- The units of momentum are kilograms metres per second (kg m/s).

Eg. A tennis ball has a mass of 0.057kg and is served at 40m/s. Its momentum is 0.057x40=2.3kgm/s

When everything has been taken into consideration, we find that momentum is always conserved.
- This is called the principle of 'conservation of momentum'.
- It states that that the total amount of momentum in a system is always the same.
- momentum at start = momentum at end

Just like a force, momentum has a direction.
- If the objects are moving in the same direction, we add up their momentums.
- If the objects are moving in opposite directions, we subtract the smaller momentum from the larger.
- Remember that the direction is the same as the larger one was.

The calculations which use momentum can be the hardest ones in the exam, though you could get lucky.

- The most important thing to remember is that the total momentum at the start is always the same as the total momentum at the end.
- If nothing is moving, that means that the total momentum is zero (because the velocities are all zero).

Eg. The tennis ball described in the previous example hits a stationary basketball (mass 0.60kg) which moves off at 2.0m/s.

tennis ball   basketball

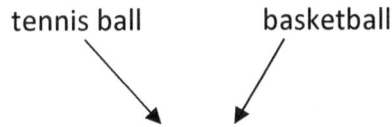

momentum at start = 2.3 + 0 = 2.3kgm/s
So, we know that the total momentum at the start is 2.3kgm/s. The basketball wasn't moving, so didn't add anything.

momentum at end = 2.3kgm/s   *(it must be the same as at the start)*
So, we know that the total momentum after the collision is 2.3kgm/s. No surprises there – it is always conserved.

Now we can do the calculations for afterwards:
momentum of tennis ball + momentum of basketball = 2.3
momentum of tennis ball + 0.60 x 2 = 2.3   *(putting in the numbers for the basketball)*
momentum of tennis ball + 1.2 = 2.3
momentum of tennis ball = 2.3 - 1.2 = 1.1kgm/s (by *rearranging the equation*)
So, we know the final momentum of the tennis ball is 1.1kgm/s. We are almost there.

Finally, we can work out the velocity of the tennis ball:
0.057 x velocity of tennis ball = 1.1 *(putting in the numbers for the tennis ball)*
velocity of tennis ball = 1.1 / 0.057 = 18m/s

So, we know that the final velocity of the tennis ball is 18m/s. That is the calculations finished. If you can do that without guidance then you are doing very well indeed!

When two objects collide with each other, the force on each object is determined by the change in its momentum and the time taken for the collision.
- The larger the change in momentum that occurs, the larger the force that is experienced. That is why the faster you hit something, the more damage is done.
- The shorter the time taken for the change in momentum to occur, the larger the force that is experienced. That is why hitting something hard hurts more than hitting something soft.

Many car safety systems, like air bags, seatbelts and crumple zones, rely on this principle.
- They cause your head to slow down over a longer period of time than if it hit the steering wheel.
- This means a smaller force is exerted on your head.
- Less force exerted means less damage done.

The same idea can be used to explain why you should bend your knees when landing from a height.
- It takes longer for your body to slow down.
- The force applied to your bones is less.
- You are less likely to damage your spine or knees.

P2.3.1 Static Electricity

We have all encountered electricity. Most people have probably felt a small electric shock (hopefully nothing too large). You have probably also played with static electricity, perhaps rubbing a balloon on your head.

If a pair of different insulators are rubbed together, for instance a plastic rod and a cloth, they both become charged.
- This is because electrons are transferred from one object to the other.
- Electrons are tiny, negatively charged particles in atoms.
- The object that gains electrons becomes negative (it gains negative charges).
- The object that loses electrons becomes positive (it loses negative charges).
- The amount of positive charge gained by one insulator is always the same as the amount of negative charge gained by the other insulator.

Charged objects exert an electrostatic force on each other.
- This allows them to push or pull each other without touching.
- Positive and negative charges will pull together.
- Positive and positive charges will push apart.
- Negative and negative charges will push apart.
- Remember – opposites attract.

Some materials hold onto their charge, but others let go of it the moment they are touched.
- Electrons can't flow through insulator (like plastic) easily, so they keep their charge.
- Electrons can flow through conductors (like metals) easily, so they lose their charge as it flows to other things.
- That is why you can't charge metal by rubbing it.

P2.3.2 Electrical Circuits

Flowing electricity can be used to do work. It is the kind of electricity we use in circuits and runs everything from lighting to computers.

We call a flow of something a 'current'.
- Electric current takes the form of a flow of electrons.
- The electrons flow from the negative end to the positive end of a cell or battery or power supply.
- The amount of electrical current is determined by the rate of flow of charge.

$$current\ (A) = \frac{charge\ (C)}{time\ (s)} \qquad I = {}^{Q}/_{t}$$

- Current is measured in amps (A).
- Charge is measured in coulombs (C).
- Time, as usual, is measured in seconds (s).

Eg. If 18C of charge flows in 3s, the current is:

$\frac{18}{3} = 6A$

We call the difference in energy for each coulomb of charge between two points the 'potential difference'.
- It is the work done by each unit of charge.
- We often call potential difference 'voltage'.
- A battery causes the energy carried by each coulomb of charge to rise.
- A resistor causes the energy carried by each coulomb of charge to drop.

$$potential\ difference\ (V) = \frac{work\ done\ (J)}{charge\ (C)} \qquad V = {}^{W}/_{Q}$$

-

Here are the circuit symbols you need to know:

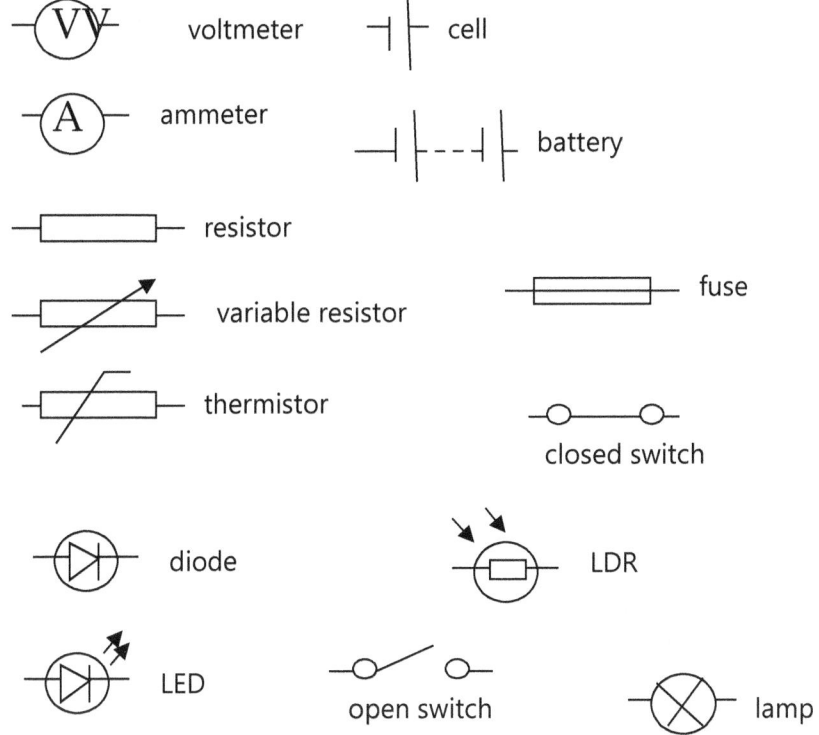

We call the opposition to the flow of current 'resistance'. It is a bit like friction for electricity.

- Resistance is caused by positive ions in the material getting in the way of the electrons that are trying to move through it. Think about how hard it is to get through a crowded corridor.
- The resistance of most materials increases as the temperature increases.
- As the material gets hotter the ions vibrate more and are more likely to get in the way of the electrons.
- In a series circuit, the resistances of the components add up simply.
    Eg. a 5Ω resistor and a 7Ω resistor have a total resistance of 12Ω.

To calculate the resistance of a component, you need to measure the current through it and the potential difference across it. Here is a circuit to do that:

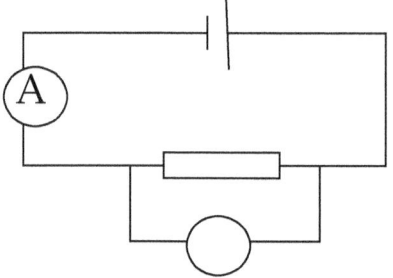

You can then use Ohm's Law:

potential difference (V) = current (A) x resistance (Ω)

$V = I \times R$

- The current through a component is directly proportional to the potential difference across it.
  ie. If you double the potential difference, you double the current (as long as the resistance stays the same).
- The current through a component is inversely proportional to its resistance.
  ie. If you double the resistance, you halve the current (as long as the potential difference doesn't change).
- Potential difference is measured in volts (V).
- Resistance is measured in ohms (Ω).

A series circuit is one which only has one loop. That means that there is only one path for current to follow.

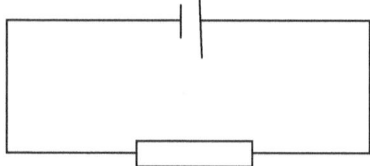

- If you put ammeters in different parts of a series circuit they will all read the same.
- The current through any point in a series circuit is the same as the current through any other point.

A parallel circuit is one with more than one loop. That means that there are several paths for current to take.

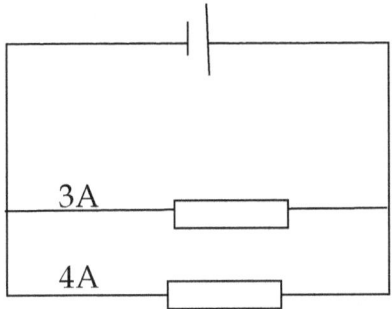

- In a parallel circuit, the currents through each branch of the circuit add up to the current through the supply.
  Eg. In the circuit above, the current through the cell is 3+4=7A.

Cells give the electrons energy to move round the circuit.
- When two or more cells in series are pointing in the same direction, the potential differences add up.

Eg. Three 1.5V cells in series provide 4.5V.

- When some cells are pointing in opposite directions, they cancel out.

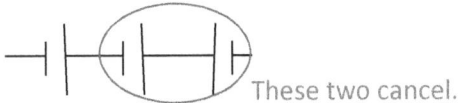

These two cancel.

You need to be able to work out how potential difference is shared out round a circuit.

- In a series circuit, the potential differences across the components add up to the potential difference across the supply.

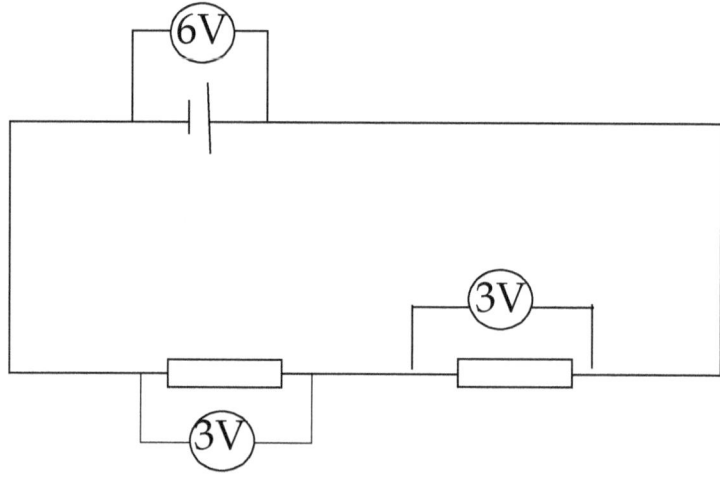

- In a parallel circuit, the potential difference across each arm is the same as across the supply.

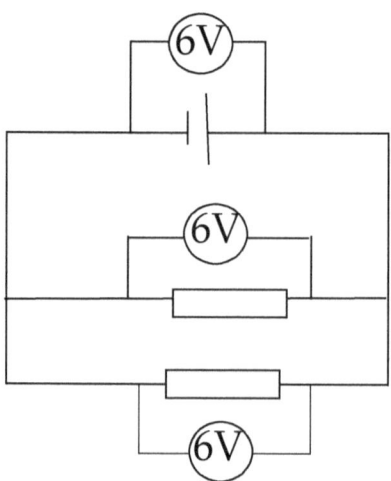

- On each branch of a parallel circuit, the potential difference across each component adds up to the potential difference across the whole branch.

A current-voltage graph (also known as an IV graph) is a graph of current (on the y-axis) against potential difference (on the x-axis).
- V stands for voltage.
- I stands for current.
- You can find the resistance at a point by reading off the voltage and the current then using the formula R=V/I.

The IV graph for an ohmic resistor is a straight line.

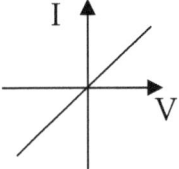

- Most metals are ohmic resistors – it just means that they obey Ohm's Law.
- As long as the temperature remains constant, the resistance of an ohmic resistor stays the same.
- You might see them called 'ohmic conductors', it is the same thing.

The IV graph for a filament lamp is a curve that looks like an italic S.

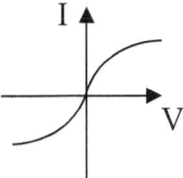

- Filament bulbs are old-fashioned lamps that work by the current heating a thin piece of wire (the filament) until it glows.
- The graph levels off because the temperature increases as more current flows through it.
- This makes the atoms vibrate more and block the movement of the electrons (ie. increases the resistance).

The IV graph for a diode is flat then shoots up at low positive voltage.

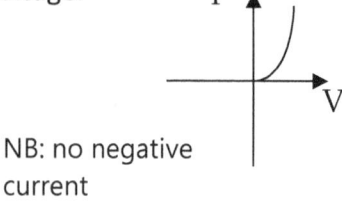

NB: no negative current

- Diodes are designed to only allow current to flow one way.
- They have very little resistance when current flows the right way.

Light Emitting Diodes (LEDs) are diodes that emit light when current flows in a forward direction.
- LEDs use a much smaller current than filament bulbs.
- They waste less energy as heat than filament bulbs.
- This means they cost less to run.
- That is why they are frequently used for standby lights on things like televisions.

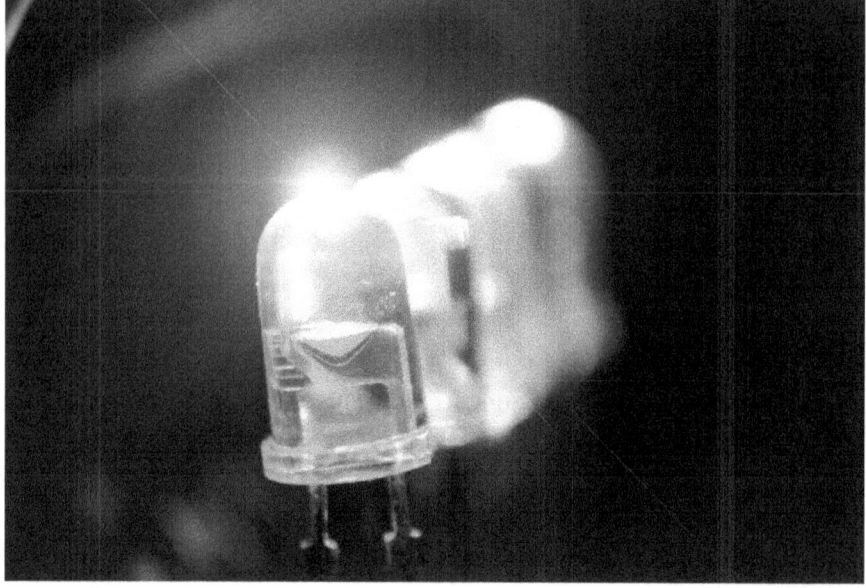

Thermistors are temperature-sensitive components.

- They have a high resistance when it is cold and a lower resistance when they are warm.
- This is the opposite of other components.
- Thermistors are often used in temperature sensors, for instance a central heating thermostat or an oven thermometer.

Light Dependent Resistors (LDRs) are light-sensitive components.

- They have a high resistance when it is in the dark and a lower resistance when it is bright.
- LDRs are often used in light meters, for instance to find a good place to grow plants, lights that automatically come on in the dark or umpires deciding when to stop play in a cricket match.

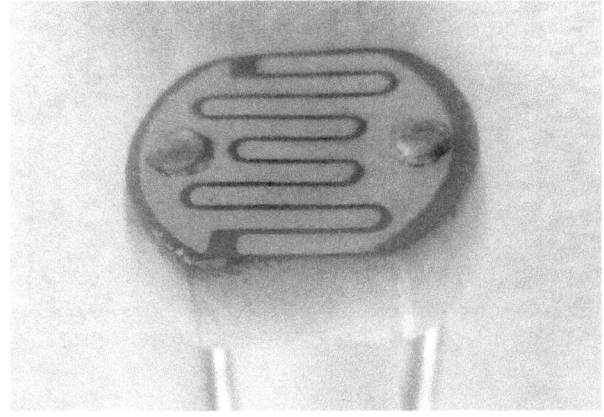

P2.4.1 Household Electricity

We all use electricity all the time, whether it is to charge our mobile 'phones or cook the dinner. We need to be able to use it safely. If nothing else, you need to know about the kind of electricity you will meet in your house!

Current that only flows one way is called 'direct current'.
- Direct current is abbreviated to 'dc'.
- Direct current is supplied by cells (and batteries). If a device only works when it is plugged in, it is unlikely to be using direct current.

Current that keeps swapping direction is called 'alternating current'.
- Alternating current is abbreviated to ac.
- Alternating current is supplied by the mains. If a device only works when it is plugged in, it is likely to be using alternating current.

If you looked at an ac current using an oscilloscope it would look something like this:

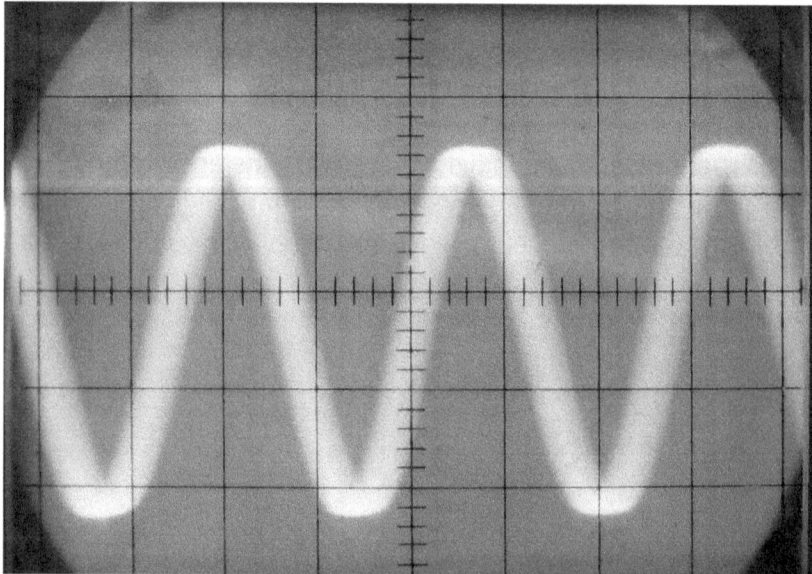

- You can read the peak voltage off the graph – it is the highest point.

- You can read the time period off the graph – the horizontal axis is time, so just see how long it is between peaks.
- You can work out the frequency by using:

$$frequency\ (Hz) = \frac{1}{time\ period\ (s)}$$

- Frequency is measured in hertz (Hz).

Different countries have different electrical systems.
- Mains voltage in the UK is 230V.
- Mains frequency is 50Hz. That means it changes direction and back 50 times a second.
- If you take a device abroad, you will probably need an adapter to change their mains voltage and frequency into ours, so that your device works without being damaged.

In the UK, most electrical devices are connected to the mains using a three-core cable and three pin plug.
- Three-core cables have three wires in the core. Obvious really!
- Each wire is coated in rubber to insulate it (so electricity doesn't escape).
- The bundle of three wires is also coated in rubber, for extra insulation and protection.

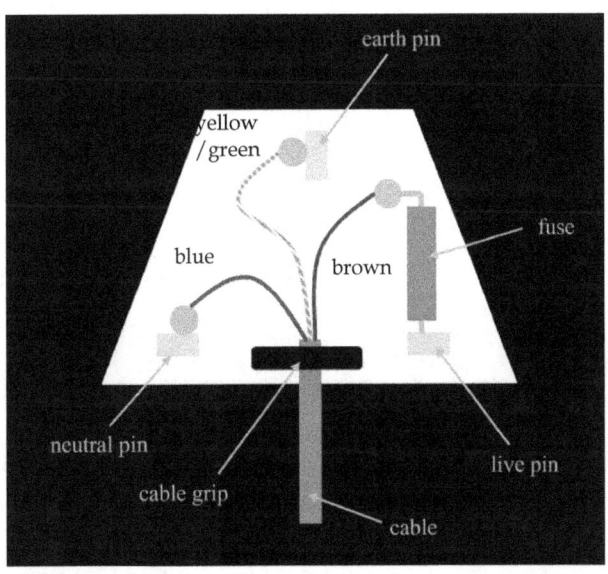

The three wires are called 'live', 'neutral' and 'earth'.
- The live wire is coated in brown rubber.
  You can think of it as the wire that the current comes in through.
- The neutral wire is coated in blue rubber.
  You can think of it as the wire that the current goes out through.
- The earth wire is coated in yellow and green striped rubber.
  It is a safety feature and it connects the outer metal case of the device to the ground (hence the name 'earth').

The fuse is a safety feature that protects against fire.
- It is made of a thin piece of wire inside a protective case.
- If there is a problem with the device or the supply and the current gets too high, the wire in the fuse melts.
- This breaks the circuit and stops the device overheating.

Often when a device stops working it is just because the fuse has 'blown'. Being able to change the fuse on a stereo could save a party!
- Some plugs will let you pop the fuse out without unscrewing the whole plug. Other types require you to take the lid off the plug to get at it.
- You need to choose a fuse that has a rating just over the normal working current of the device. Eg. Choose a 3A fuse for something that normally uses 2.4A.
- If the fuse keeps blowing there may be a deeper problem with the device. You should stop using it and get it looked at by an expert.

Some devices, such as room heaters and ovens, need a very large current to be able to work.
- They need thick cables to handle the large current.
- They need fuses with high ratings so they don't blow in normal use.
- So, if a device has a thick cable, it is likely to need a large fuse.

Circuit breakers can prevent fire, just like fuses.

- Circuit breakers can be reset, but fuses need replacing once they have 'blown'.
- This means one benefit of a circuit breaker is that it can be tested periodically, to ensure it is still working.

If a live wire comes loose inside a device it can touch the outer case.

- If the outer case were made of metal, it would become live.
- If the fault occurs whilst someone is actually touching the device, the fuse doesn't blow fast enough to prevent them getting electrocuted.

The earth wire allows the current from the outer case to flow safely to the ground.

- The earth wire has a very low resistance, so the current will be very high.
- This will blow the fuse, and thus cut off the supply of electricity.
- The earth wire and the fuse together prevent electrocution, but only if the device malfunctions whilst not being held.

Residual current circuit breakers (RCCBs) react quickly enough to prevent the user getting electrocuted.

- They work by comparing the current going in the live to the current coming out the neutral.
- If the currents are the same then everything is OK.
- If the current coming out is less than the current going in then some is going somewhere it shouldn't.
- If that is the case, the RCCB breaks the circuit.

We call devices coated in rubber or plastic 'double-insulated'.

- The outer insulation on the device is first layer and the insulation on the wires is the second.
- With a double-insulated device, there is no chance of the user touching a live part, and so there is no need for an earth wire.
- In that case, two-core cables can be used.
- Two-core cables only contain live and neutral wires.

P2.4.2 Current, Charge and Power

It is often useful to know some things about devices in your home, like current they draw and the power they use. The current lets you know what fuse it needs. The power tells you the rate at which it uses energy, and so how costly it will be to run.

We call the rate of transferring energy 'power'.

- $$power\ (W) = \frac{energy\ transferred\ (J)}{time\ taken\ (s)}$$

The power outputs can be split into two categories.

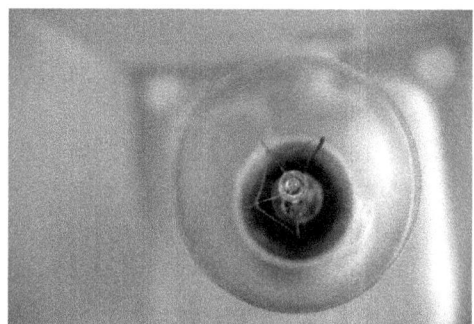

- Useful output – doing what the device is designed to do.
- Wasted output – everything else.
- A common waste is heat.
- Remember that the total output is equal to the total input.

How much of the power is wasted is one factor that people consider when choosing which device to buy.

For example, old-fashioned filament bulbs wasted a lot more energy as heat than modern energy-saving light bulbs (like compact fluorescent lamps or LEDs).

You need to be able to work out things like the power your device will use, or its working current.

- *power (W) = voltage (V) x current (A)* $P = V \times I$
- If the device doesn't have the current written on it, it will state the power and voltage. You can use the formula above to calculate the current.
- Once you know the current you can use that information to choose the right fuse.
- The right fuse is the one just a bit bigger than the current the device is supposed to use.

- That means it won't blow every time the device is switched on, but will cut out with the minimum increase in current above normal.

Eg. My hairdryer has a power of 1000W and runs off 230V mains.

Its working current is 1000/230 = 4.3A.

Given a choice of 1A, 3A, 5A and 13A fuses I would choose 5A.

The potential difference across a component and the charge that flows through it determine the energy transferred by a component

- energy transferred (J) = potential difference (V) × charge (C)

$$E = V \times Q$$

- Eg. If 4.0C flows from a 1.5V cell 4.0x1.5=6.0J of energy is transferred.

When you come to choose a new appliance or device you should think about:

- the voltage it will need
- whether it needs ac or dc
- the power it will draw
- the current it will draw
- how efficient it will be.

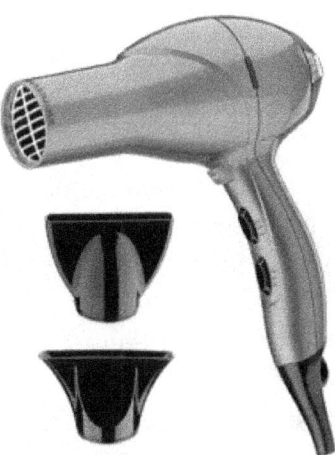

P2.5.1 Atomic Structure

All objects are made up of atoms. There have been many theories over time about what matter is, but is only relatively recently that people really started to understand atoms.

Atoms are the smallest bits of an element you can get.
- Atoms are mostly empty space.
- At the centre of every atom is a tiny, dense ball.
- We call this ball the 'nucleus'.
- The nucleus is roughly 100000 times smaller than the atom.
- The nucleus contains only protons and neutrons.
- The nucleus is surrounded by a cloud of electrons.

Before 1909 no-one even suspected the existence of the nucleus.
- Electrons had been discovered, but scientists were unsure about the structure of the rest of the atom.
- One theory was calle the 'Plum Pudding Model'. This said that the negative electrons were held in a solid positive 'batter'.

In 1909 Geiger and Marsden did an experiment to study the structure of atoms.
- They fired alpha particles at gold atoms, so it is sometimes called 'the alpha scattering experiment'.
- Most alpha particles went straight through. This proved that atoms were mostly empty space.
- Some alpha particles were deflected a bit. This showed there was a concentration of positive charge in each atom.
- A few alpha particles bounced back. This showed there was something small with a lot of mass, which came to be called the nucleus.
- This evidence meant that the Plum Pudding Model had to be discarded in favour of the Nuclear Model.
- They were supervised by Rutherford, so it is sometimes called the 'Rutherford scattering experiment'.

The numbers involved with atoms are often ridiculously small. This can make them hard to work with so we tend to use 'relative' values as a way to simplify things.

particle	relative mass	relative charge
proton	1	+1
neutron	1	0
electron	1/1836	-1

- Every nucleus of a particular element has the same number of protons.
- We call the number of protons in an atom the 'atomic number'.
- The number of protons + number of neutrons in an atom is called its mass number.

Eg. A nucleus with a mass number of 7 and an atomic number of 4 has 3 protons and 7-4=3 neutrons.

Atoms are always neutral.
- That means that they have no overall electric charge.
- That is because they always have the same number of electrons as protons.
- If an atom loses one, or more, electrons it becomes a positive ion.
- If an atom gains one, or more, electrons it becomes a negative ion.
- The number and arrangement of electrons is what gives particular chemical properties.

Atoms of a particular element can still have different numbers of neutrons.
- These are called isotopes.
- All the isotopes of an element have the same chemical properties.
- The only differences are their mass and how they decay (see next chapter).

P2.5.2 Atoms and Radiation

The nuclei of atoms can change, decaying into different elements. In the process they release harmful, but also useful, radiation.

Some nuclei are stable.
- They don't change by themselves.
- The only way to make them change is to hit them with something like a neutron.

Some nuclei are unstable.
- They change by themselves.
- We call this change a 'decay'.
- When they decay, they give off radiation.
- We say they are 'radioactive'.
- The radiation is given off randomly.
- It is impossible to predict when a particular nucleus will decay.

Nuclei can give off three main types of radiation.
- Alpha particles – made of two protons and two neutrons (the same as a helium nucleus).
- Beta particles – fast electrons.
- Gamma radiation – an electromagnetic wave.
- However, gamma cannot be released on its own – it only ever follows an alpha or a beta emission.

The types of radiation emitted by a source can be identified by studying their penetrating power.
- Alpha is stopped by only a few centimetres of air or a sheet of paper (or skin, but that isn't recommended for an experiment).
- Beta needs a few tens of centimetres of air or a few millimetres of metal (such as aluminium) to stop it.
- Gamma can penetrate many metres of air or several centimetres of lead.
- Eg. If a source emits something that can penetrate a sheet of paper but not a cm of lead then it is emitting beta radiation.

This variation in penetration is due to the differences in ionizing power.

- Alpha is the most strongly ionizing, so it rapidly loses its energy to atoms in the material.
- Gamma is a least strongly ionizing, so it can travel a long way through a material without losing much energy.

We can also use magnetic fields to determine which kind of radiation a source is emitting.

- Alpha and beta are deflected by magnetic fields.
- This is because they are charged.
- They are deflected in opposite directions to each other.
- This is because they have opposite charges.
- Beta particles are deflected through larger angles than alpha particles.
- This is because alpha particles have more mass and so need a bigger change in momentum.

We can write a symbol that contains all the information about a nucleus.

- $^{M}_{A}X$
- X is a letter or letters that represents the name of the element.
- A is the atomic number, the number of protons in the nucleus.
- M is the mass number, the number of protons and neutrons in the nucleus.

Eg. $^{14}_{6}C$ is an isotope of carbon. It has a mass number of 14 and an atomic number of 6.

It is often useful to write an equation to represent a decay.

Eg. $^{14}_{6}C \rightarrow ^{14}_{7}N + ^{0}_{-1}\beta$

- The numbers on the top row need to add up to the same number on both sides of the equation.
- The numbers on the bottom row need to add up to the same number on both sides of the equation.

We use the phrase 'half life' to describe how quickly an unstable isotope decays.

- The half life of an isotope is the average time it takes for half the radioactive nuclei to decay.
- It is also the average time it takes for the radioactivity of the sample to halve.
- The shorter the half life, the faster the isotope decays.
- A large sample is required to measure the half life accurately, because radioactivity is a random process.

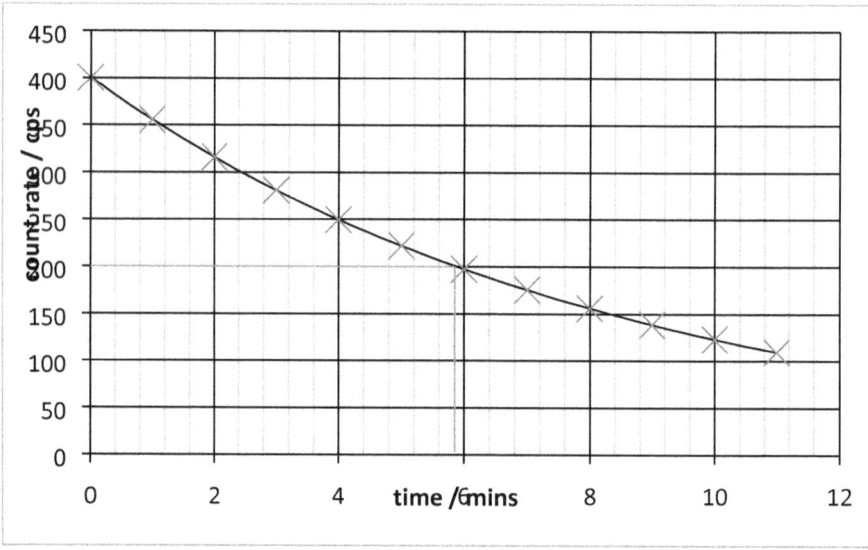

In the example shown in the graph above, the initial count rate was 400cps. Half of that is 200cps, so we read across at 200cps until we meet the curve and then down from the curve to the x-axis.

We find that the half life is 5.8 minutes.

Radioactive isotopes have many uses, such as:

- Carbon dating – for finding the age of samples in archaeology.
- Medical tracers – to diagnose certain illnesses, such as a hyperactive thyroid.
- Radiotherapy – for treating some forms of cancer.

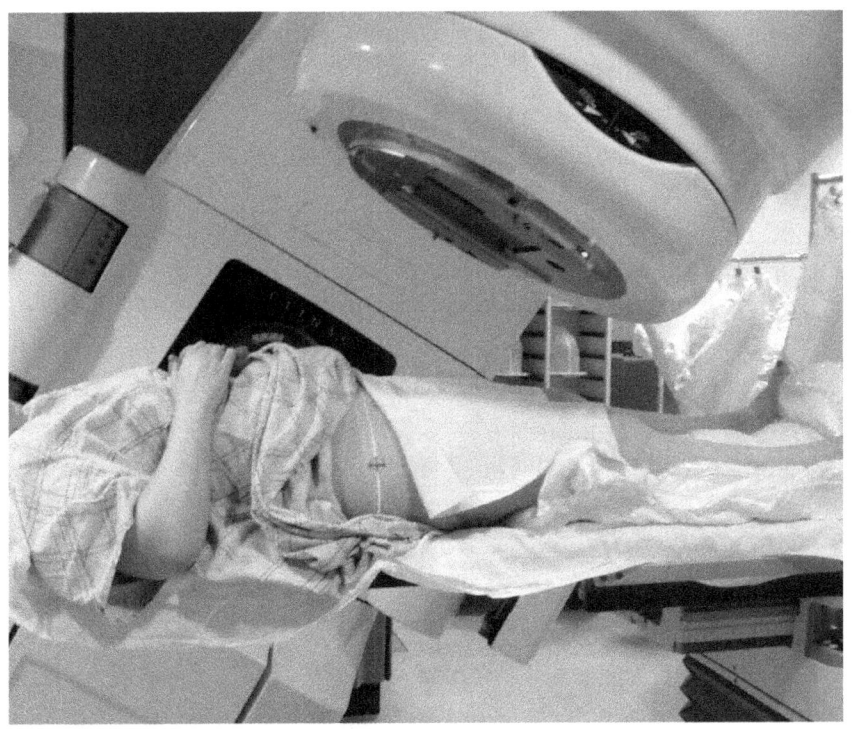

When you choose what isotope to use, the first thing you need to decide on is how much they need to penetrate.

- For tracers, the radiation needs to be able to escape and reach the detector.

  In most cases this means they need to be gamma emitters.

- A classic example of an industrial use of radioactive isotopes is a paper factory.

  The thickness of the paper is measured by how much radiation can get through it.

  Alpha is the most sensible to use because neither beta nor gamma will be stopped by even a thick sheet of paper.

- Radiotherapy needs the cancer cells to be destroyed inside the body, so gamma is frequently used.

  Some radiotherapy uses beta sources surgically implanted in the patient.

After choosing alpha, beta or gamma, you need to consider the half-life of your isotope.

- The isotope cannot 'run out' during use.
- However you don't want it hanging around too long afterwards.

Eg. A medical tracer needs to emit long enough to trace the problem, but if its half-life is too long the patient will be exposed for an unsafe length of time.

Exposure to radiation can cause harm, such as:

- radiation burns
- acute radiation sickness
- cancer
- sterility
- genetic mutations.

Alpha is safe if kept at a distance, as it can only penetrate a few centimetres of air. However it is very dangerous if it gets inside the body, eg. inhaled, as it does a lot of damage to soft tissue.

Beta is not as damaging as the others, so long as you are not exposed for too long.

Gamma can penetrate your whole body and has high energy so can do a lot of damage.

We are exposed to radiation all the time.

- This is called 'background radiation'.
- The amount you are exposed to depends on where you live and what your job is.
- When doing an experiment measuring radiation you must take away the background count from every reading.
- We call the new value the 'corrected count rate' or 'ccr'.

Eg. I record 35 counts per minute. The background in that location is 4 counts per minute. The radioactivity of the source in my experiment is 35-4 = 31 counts per minute.

Most background radiation comes from natural sources such as:
- rocks
- the Sun
- cosmic rays
- gases in the air
- food and drink.

Some comes from man-made sources of radiation such as:
- hospitals
- nuclear weapons tests
- nuclear accidents.

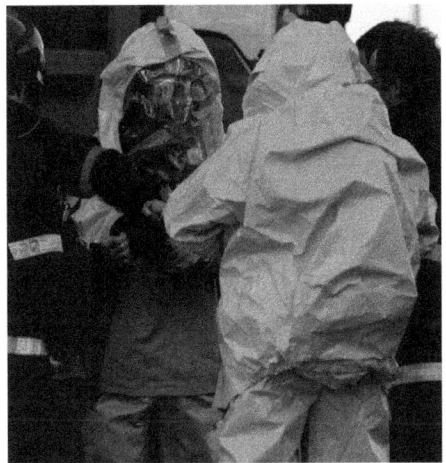

People exposed to high levels of radiation need to take steps to bring the risk down.
- The exposure could be from their job, like a hospital worker, or from where they live, due to the types of rocks underground.
- The steps for a worker could include limiting the time they are exposed and/or wearing protective clothing (perhaps lined with lead). This could, however, hinder how effectively they can do their job.
- The most common radiological risk at home is radon gas.
- Modern houses in areas with high background are designed to prevent radon gas building up in them.
- This is often by including good ventilation – air bricks, window vents and fans.

P2.6.1 Nuclear Fission

Nuclear reactors are still an important source of electricity for our country. Some people think they are vital; others that building more would be a disaster. Understanding how they work wil help you make up your own mind.

Nuclear fission is another way a nucleus can decay.
- Fission is when a heavy nucleus splits into two lighter nuclei.
- We call these lighter nuclei the 'daughters'.
- The split is triggered by the nucleus absorbing a neutron.
- The process also releases two or three neutrons.
- The fission products have a lot of kinetic energy.

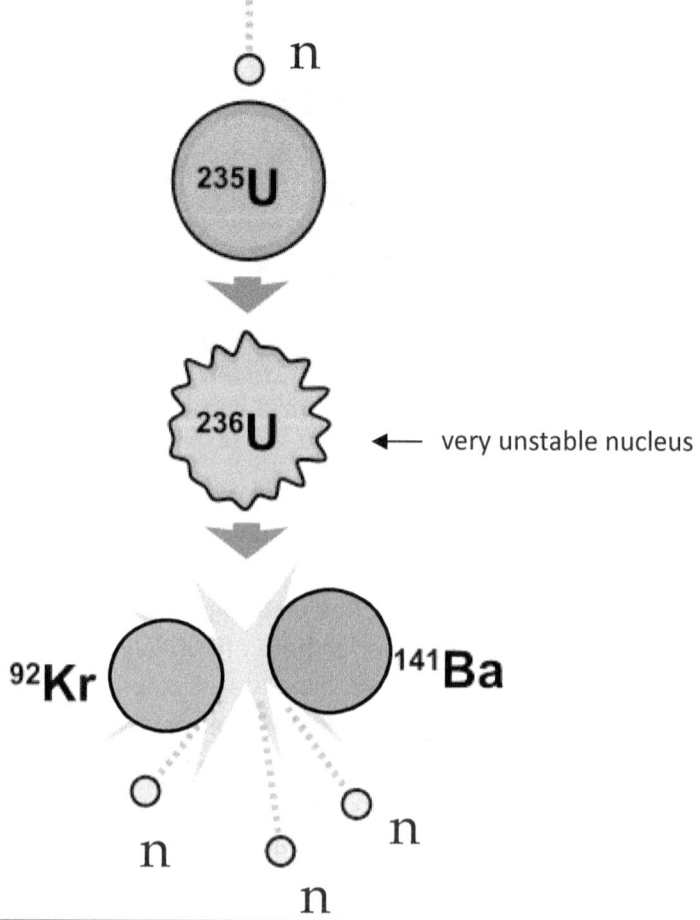

^{235}U

n

^{236}U ← very unstable nucleus

^{92}Kr ^{141}Ba

n   n   n

If there are enough heavy nuclei in the sample, the neutrons released by a fission event can go on to hit other heavy nuclei and cause them to undergo fission.

- We call this process a 'chain reaction'.
- If the chain reaction runs away, it can cause a large explosion, like in an atomic bomb.

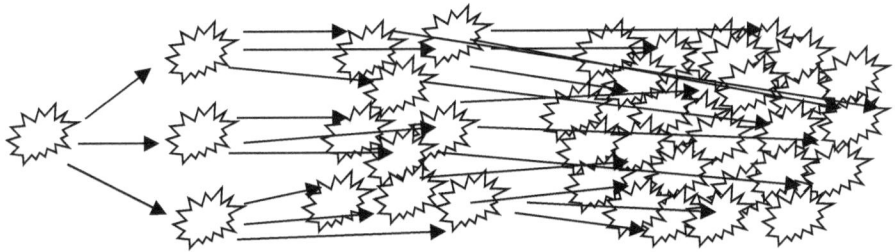

- If the chain reaction is controlled, it can be used to generate a steady output of energy, like in a nuclear power station.

- Uranium-235 and plutonium-239 are the two most common isotopes we use for fission.
- Nearly all reactors use uranium-235.

There are many pros and cons of nuclear reactors.

- Some people are against fission reactors because they produce radioactive waste that is dangerous to transport and store. They are also worried about the risks of an accident sending radioactive dust into the atmosphere.
- Others support nuclear power because it doesn't use up fossil fuels and does not emit gases that cause global warming and acid rain.

P2.6.2 Nuclear Fusion

Fusion is thought by many scientists to be the energy source of the future. It powers the Sun and other stars, but we haven't quite got it working well enough on Earth to be used outside of a laboratory.

Nuclear fusion is when two light nuclei stick together to make a heavier nucleus.
- Energy is released as a result.
- The fuel is not radioactive.
- The waste products are non-polluting.

Fusion occurs naturally in stars.
- They are hot and dense enough for the nuclei to stick together easily.
- Stars form when a cloud of dust and gas, called a nebula, collapses under its own gravity.
- This forms a protostar.
- When the protostar becomes dense and hot enough, the hydrogen nuclei in it start colliding.
- They have enough energy to start fusion.
- The outwards force from the fusion reaction balances the inwards gravitational force.
- This stops the star collapsing.

This stable phase is called a main sequence star.

- A main sequence star makes all the elements up to iron, by fusion.
- Any dust and gas left over from the formation of the star can pull together under gravity (larger masses pulling in smaller masses) to form planets, asteroids etc.
- Stars are so massive that they contain enough fuel to support fusion for millions of years.

When the star runs out of hydrogen in starts fusing heavier nuclei.

- The fate of the star depends on its mass - the more mass the faster it uses up its fuel and the sooner it runs out.

A star like our Sun will go through the following stages:

- Red Giants – stars that have run out of hydrogen and are rapidly fusing heavier elements.
- White Dwarves – hot balls of iron that have run out of fuel for fusion.
- Black Dwarves – cold balls of iron.
  They are cooled down White Dwarves and don't give off visible light.

Stars many times the mass of our Sunwill go through the following stages:

- Red Supergiant – an even larger version of a red giant.
- Supernova – an explosion which makes all the elements above iron and distributes them throughout the Universe.

They then either form:

- Neutron Stars – small balls of neutrons about the size of a city but several times the mass of our Sun.
- Black Holes – stars that are so dense they have collapsed down to a point.
  Their gravity is so strong that not even light can escape (hence the name).

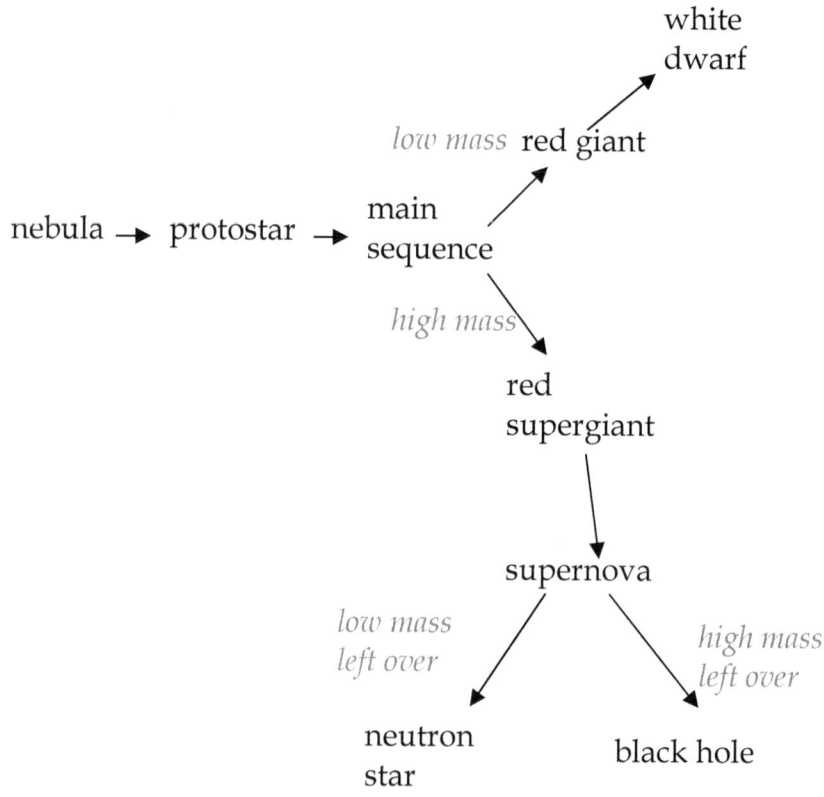

Life Cycle of a Star
High mass means much bigger than our Sun.
Low mass means similar to our Sun.

The elements that are blown off by red giants and supernovæ end up forming new nebulæ and the process begins again.

Our Sun is currently a main sequence star.
- When it becomes a red giant it will be big enough to swallow up Earth in its orbit.
- It will then be left as a white dwarf.
- Eventually it will cool to be a black dwarf.

When the first elements condensed after the Big Bang, the only element was hydrogen.

- This was because the universe was so hot that the protons were moving so fast they couldn't combine with other protons.
- All the other elements in existence were made in stars through the process of fusion (until we started using particle colliders that is).

Also by the Author

Physics
Physics Problems for GCSE
http://www.archaeoroutes.co.uk/edphys/problems.php

The Best Bits of Physics
http://www.archaeoroutes.co.uk/bbop

AQA GCSE Physics Revision Guides
http://www.archaeoroutes.co.uk/edphys/revision.php

AQA and OCR A-Level Physics Practice Tests
http://www.archaeoroutes.co.uk/edphys/exams.php

Science Fiction
Two Democracies - Independence
http://lrd.to/xZJGd4yPXo

Archaeology and Walking
Walking Through the Past series
http://www.archaeoroutes.co.uk

Credits

The following images are licensed under a Creative Commons Attribution ShareAlike Unported 3.0 license (https://creativecommons.org/licenses/by-sa/3.0/deed.en):

p38 – https://en.wikipedia.org/wiki/Light-emitting_diode#/media/File:Orange_LED_emitting.png

p39 – https://en.wikipedia.org/wiki/Photoresistor#/media/File:LDR_1480405_6_7_HDR_Enhancer_1.jpg

p52 – blazingfires13 - http://blazingfires13.deviantart.com/art/Radioactive-sign-brick-manip-1-368550062

p53 – heb@wikimediacommons - https://en.wikipedia.org/wiki/Hazmat_suit#/media/File:MHE_-_KBH_Brandvaesen_-_HAZMAT_3a.jpg

p56 – Mike Garrett - https://en.wikipedia.org/wiki/Divertor#/media/File:Alcator_C-Mod_Tokamak_Interior.jpg

The following images are licenced under a Creative Commons Attribution Generic 2.0 license (https://creativecommons.org/licenses/by/2.0/):

p39 - Tomi Knuutila - https://www.flickr.com/photos/yourbartender/5447374145/in/photostream/

p40 – Tess Watson - https://www.flickr.com/photos/tessawatson/427116107/in/gallery-toolgirl-72157625760588818/

p51 – Stevenfruitsmaak - https://en.wikipedia.org/wiki/Radiation_therapy#/media/File:Radiation_therapy.jpg

The following images are licenced under a Creative Commons Attribution ShareAlike Generic 2.0 license (https://creativecommons.org/licenses/by-sa/2.0/):

p45 – Kalita Kabir - https://www.flickr.com/photos/irsein/5225483461

The following images are licenced under the Open Government License 3.0 (https://www.nationalarchives.gov.uk/doc/open-government-licence/version/3/):

p19 – Highway Code

All other images are copyright Alasdair C Shaw, have been released into the public domain by their author, or are available under a CC0 license.

www.ingramcontent.com/pod-product-compliance
Lightning Source LLC
Chambersburg PA
CBHW071633170526
45166CB00003B/1307